# How Contagion Works

Paolo Giordano is a physicist and an internationally bestselling author. His first novel, *The Solitude of Prime Numbers*, was translated into more than 40 languages worldwide and won the Premio Strega (the Italian Booker Prize).

Paolo is the author of the forthcoming novel *Heaven and Earth* and the creator of the new HBO series *We Are Who We Are*, co-written and directed by Luca Guadagnino.

He holds a master's degree and a PhD in theoretical physics. He lives in Rome, Italy.

By Paolo Giordano

*The Solitude of Prime Numbers*
*The Human Body*
*Like Family*
*Heaven and Earth*

*How Contagion Works*

# How
# Contagion Works

Science, Awareness and Community
in Times of Global Crises

## Paolo Giordano

Translated from the Italian by Alex Valente

WEIDENFELD & NICOLSON

First published in Great Britain in 2020 by Weidenfeld & Nicolson
an imprint of The Orion Publishing Group Ltd
Carmelite House, 50 Victoria Embankment
London EC4Y ODZ

An Hachette UK Company

1 3 5 7 9 10 8 6 4 2

A CIP catalogue record for this book is
available from the British Library.

ISBN (Paperback) 978 1 4746 1928 8
ISBN (eBook) 978 1 4746 1929 5

Typeset by Born Group
Printed and bound in Great Britain by Clays Ltd, Elcograf S.p.A.

www.orionbooks.co.uk
www.weidenfeldandnicolson.co.uk

# How
# Contagion Works

# Grounded

The Covid-19 epidemic is set to be the most significant health emergency of our time. Not the first, not the last, maybe not even the most horrific. Most likely in the end the death toll won't be any higher than other illnesses – and yet, three months since its debut, it has reached its first milestone: Sars-Cov-2 is the first virus to spread this quickly on a global scale. Many others similar in nature, such as its predecessor Sars-Cov, have been quickly dealt with. Some, such as HIV, have been hiding in the shadows for years. Sars-Cov-2 had more gumption. And its boldness has revealed something we had long known but had been unable to measure with precision until now: the multiple levels and layers that connect us to each other, everywhere, and the complexity of the world we inhabit – its social, political, financial motives and its interpersonal and psychological structures, too.

I'm writing this on a rare February 29th, a Saturday of this leap year. The confirmed global cases of infection have surpassed the 85,000 threshold – almost 80,000 in China alone – and the death toll is around 3,000. These numbers have been my silent companions for the past month. Even now, the Johns Hopkins University interactive map is open in front of me. The areas of infection are identified by red circles in stark contrast against the grey background. Perhaps another choice of colours might have been better, less alarming, but we all know how this works: viruses are red, emergencies are red. China and Southeast Asia have disappeared beneath a giant red stain, but the entire world is pockmarked, and the rash is bound to get worse.

Italy, much to everyone's surprise, has also found itself leading the race in this anxious competition. This has happened entirely by chance: in a few days – suddenly, unexpectedly – other countries could find themselves in even worse conditions. At this moment of crisis, the expression 'in Italy' has no meaning: there are no more borders, regions, neighbourhoods. The nature of what we're going through is above identities and cultures. The epidemic is the ultimate proof of how our world has become globalised, interconnected, inextricable.

I know all this and yet, as I watch the red disc growing over Italy, I can't help but feel troubled by it, as we all are. My bookings for the next few days have been cancelled because of the containment procedures – some I called off personally – and I found myself surrounded by an unexpected emptiness. A predicament shared by many: we're living through a suspension of daily activities and routines, a pause in the usual rhythm of our lives – like one of those songs where the drums stop abruptly and the music seems to expand in the emptiness left behind. Schools are closed, very few planes are moving across the sky, solitary footsteps echo in museum corridors. Everywhere is shrouded in more silence than usual.

I decided to make use of this void by writing. Writing can sometimes be an anchor that helps us stay grounded and hold back fear. But there is also another reason: I don't want to lose what the epidemic is revealing about ourselves. Once the emergency is over, any temporary awareness will also disappear – that is the nature of illnesses.

As you read these pages, the situation will have changed. The numbers will be different, the epidemic will have spread further: it will either have reached every corner of the globe, or it will have been stopped, but some of the reflections emerging from the contagion will still apply. Because what is happening isn't a random accident or a scourge. And it's nothing new, either: it has happened before, and it will happen again.

# Nerdy afternoons

I remember afternoons back in my first years of high school, when I'd spend all my time simplifying expressions. I'd copy a long line of symbols out of the textbook, then reduce it to a concise, understandable result: $0$, $-\frac{1}{2}$, $a^2$. The world outside of the window would darken and the landscape would give way to the reflection of my face lit up by my desk lamp. Those were peaceful afternoons. Bubbles of order at an age when everything inside and outside of me – especially inside – seemed to tend towards chaos.

Long before writing, maths was my trick to keep anxiety at bay. To this day, there are mornings when I wake up and I improvise calculations and numeric sequences; it's usually a symptom that something's wrong. I suppose all of this makes me a 'nerd' and I don't mind. Plus, these days it seems that maths isn't just a nerdy pastime after all, but rather the indispensable tool in understanding what is happening to us and keeping our irrational tendencies in check.

Epidemics are mathematical emergencies first and foremost. Because maths isn't the science of numbers – not really – it's the science of relations: it describes the bonds and the exchanges between different entities, regardless of what these entities might be made of, abstracting them into letters, functions, vectors, points and planes.

The contagion is an infection of our relations.

# Contagion by numbers

We could see it gathering like dark storm clouds on the horizon, but China is far away and 'anyway, stuff like that would never happen here'.

When the contagion arrived, in full force, it left us all stunned.

To find a way through the disbelief I turned to maths, starting with the SIR model, the invisible bone structure of every epidemic.

A notable distinction: Sars–Cov-2 is the virus, Covid-19 the illness. They have unfriendly, impersonal names, maybe chosen to limit their emotional impact, but they are more precise than the popular 'coronavirus'. So I will be using them instead. To simplify even further, and to avoid misunderstandings with the 2003 Sars contagion, I'll be abbreviating Sars–Cov-2 to Cov-2 throughout.

Cov-2 is the most elementary form of life we know. To understand its actions, we need to inhabit its dumb intelligence, look at it like it looks at us. And remember that Cov-2 doesn't care about us, our age, gender, nationality, personal preferences. The entire human species, in the eyes of the virus, falls into one of three categories: the Susceptible, those it can still infect; the Infected, those it already has; the Recovered, those it can no longer infect.

Susceptible, Infected, Recovered: SIR.

According to the contagion map pulsating on my screen, the Infected around the world are currently 40,000; the Recovered, a little higher. The group we need to keep an eye on, however, is the third, the one that is never reported.

The Susceptible to Cov-2, the humans that can still contract the virus, are 7.5 billion, give or take a handful.

# R nought

Let's pretend there are 7.5 billion marbles, representing the world's population. We are susceptible and perfectly still, when suddenly an infected marble rolls at full speed against us. That infected marble is patient zero and hits two more marbles before coming to a halt. The latter two bounce off and hit two more each. And so on.

And again.

And again.

That's how contagion starts, as a chain reaction. During its first phase it grows in a way that mathematicians call exponential: more and more people are infected, with increasing speed. The speed depends on a number, the hidden heart of each epidemic: it's noted with the symbol $Ro$ – pronounced 'r-nought' – and each illness has its own. In the marbles example, $Ro$ was exactly two: each Infected caused an average of two Susceptibles to also contract the virus. For Covid-19, $Ro$ is approximately 2.5.

It's hard to say whether this is low or high. It doesn't even make that much sense. Measles has an $R_0$ of around 15, while the so-called 'Spanish flu' from last century was around 2.1 – but that didn't stop it from killing tens of millions of people across the world.

What matters right now is that things are going really well if $R_0$ is lower than 1, meaning that each Infected can only infect less than one other person. In this case, the spread is halted, the illness is all show, no substance. If, on the other hand, $R_0$ is higher than 1, even slightly, we have an epidemic on our hands.

The good news is that $R_0$ can change. In some ways, this is down to us. If we reduce the spread of the contagion, if we change our behaviours in order to make it harder for the virus to bounce from one person to another, $R_0$ decreases and the contagion slows down. That's why in Italy we've stopped going to the cinema. If we're able to be steadfast long enough, $R_0$ will slide under the critical value of 1 and the epidemic will grind to a halt.

Lowering $R_0$ is the mathematical reason behind our self-sacrifice.

# This crazy non-linear world

In the afternoon, I wait for the Civil Protection Agency's bulletin. It's the only thing I focus on: there are other things happening in the world – all important, still very much present on the news – but I don't care.

On 24 February, the confirmed Infected in Italy were 231. The following day, the number rose to 322, the following to 470; then 655, 888, 1,128. Today, a rainy first day of March, the number is 1,164. It's not what we want to hear and – crucially – it's not what we are programmed to expect.

To use numbers that are easier to handle, let's suppose that yesterday we had ten cases of infection, and today we have twenty. Our instinct leads us to think that tomorrow's bulletin will report thirty cases. Then ten more, then another ten. When something grows, we tend to think it's going to follow the same pattern every day. Mathematically speaking, we expect the growth to be linear. We can't help it.

This time, however, the actual rate of increase gets higher by the hour; it seems to be spiraling out of control. We could claim this was another way in which the virus caught us unprepared, but it would be giving it too much credit: in reality, nature itself doesn't follow a linear structure. Nature prefers growth that is either vertigo-inducing or decidedly softer: exponentials and logarithms appear everywhere in its equations. Nature is, by its own nature, non-linear.

Epidemics are no exception. But a behaviour that doesn't surprise scientists can shock everyone else. The increase in the number of infected people thus becomes 'an explosion', news headlines report 'worrying', 'dramatic' changes that were foreseeable from the very beginning. This distortion of what we consider to be 'normal' generates fear. The truth is that the increase in Covid-19 cases in Italy or anywhere else was never going to be linear: this phase of the contagion was always going to see a much faster growth and there is nothing – nothing – mysterious about it.

# Stopping the spread

'How do you stop something that keeps growing?'

'With incredible strength. With incredible sacrifices. With incredible patience.'

Now we know that facing the epidemic equals dragging down its $R_0$ value. It's very much like trying to repair a leaky tap without shutting off the water mains: if the pressure in the pipes is very high, we need to make sure the tap doesn't spray water in our faces before we can take care of anything else. This is the strength phase.

If $R_0$ is kept beneath the critical value long enough – a length of time during which all pre-existing cases have been identified and have been contained, and most of them have passed the infectious period – then we'll start seeing a decrease. The contagion still grows, but more slowly. This is the sacrifice phase.

When talking about $R_0$ earlier, however, I have been a bit too precipitous. There is also some bad news: at the exact moment when the emergency containment measures are suspended – be it in China or Italy – there is a high likelihood that $R_0$ will jump back to its 'natural' value of 2.5. If you take your hand off a pipe under pressure, water will start gushing back out of it. The contagion goes back to spreading exponentially.

And so, the third, hardest, phase begins: patience.

# Hoping for the best

I went to a dinner party last night. This is the last one, I told myself. Once we cross the 2,000 cases threshold, I'm self-isolating. I didn't kiss or hug anyone as I arrived, to their disappointment. This epidemic seems to have gone to my head. I'm a mild hypochondriac; I regularly ask my wife to check my temperature in the evenings, but this isn't it. I'm not afraid of getting ill.

So what is it?

I'm afraid of everything that the contagion can change. Of discovering that the structure holding up civilisation as I know it is nothing but a house of cards. I'm afraid of annihilation, but also of its opposite: that fear will eventually pass without leaving any trace of change behind.

During dinner, everyone kept saying, 'It'll be over in a few days,' and, 'Of course, just another week and everything will go back to normal.' A friend of mine asked me why I was so quiet. I shrugged, didn't answer; I didn't want to come across as an alarmist or, worse, jinx anything.

Even though we have no immunity against Cov-2, we have developed a resistance to anything that has the potential to unsettle our lives: the unknown, the disconcerting, the new and scary. We always want to know the exact date something will start and then come to an end, so that we can meticulously schedule our lives. We are used to imposing our conception of time onto nature, not the other way round. Therefore, I *demand* that the contagion end in a week, so we may return to normality.

But during contagion, we need to know what we're actually allowed to hope for. Because hoping for the best doesn't necessarily mean we're hoping for it in the best way. Waiting for the impossible to happen, or even just the highly improbable, will only bring us repeated disappointment. The flaw of wishful thinking, in a crisis of these proportions, isn't that it merely turns out to be false, but rather that it feeds into our growing sense of anxiety and fear.

# Really stopping the spread

'So how do we really stop this from spreading?'
'With a vaccine.'
'What if there is no vaccine?'
'With more patience.'
Epidemiologists know that the only way to stop the epidemic is to reduce the number of Susceptibles. Their density needs to become too low for the virus to keep spreading. We need to keep the marbles apart: when the sequence of marbles hitting each other is sufficiently low, the chain reaction will stop.

Vaccines have the mathematical power of making Susceptibles morph into Recovered without going through the illness. Vaccines matter to us because they protect us from the virus, but they matter even more to infectious disease specialists because they spare us from the epidemic. We wouldn't even need for everyone to be vaccinated; we'd just need a big enough percentage to reach what is known as 'herd immunity'.

But Cov-2 caught us unprepared and unarmed, with no antibodies or vaccines. It's too new for us. Looking at it through the SIR model, its novelty means we are all Susceptibles.

Therefore, we'll have to resist for the required length of time. Our only vaccine right now is an uncomfortable form of cautiousness.

# Cautious calculations

I wanted to get to the mountains no matter the cost. That holiday was a long-awaited reward to myself after exam season. My friends were as invested as I was and everything had already been paid for, including the hotel in Les Deux Alpes and – out of an extraordinary bout of pre-planning – our weekly ski pass. As we exited the Salbertrand tunnel, we found ourselves engulfed in a snowstorm. It had only just started, the roads were still clear. We told each other: we can still make it. Another few miles and the traffic came to a stop. We put on snow chains – which took some effort, especially for a group of inexperienced young drivers – and by the time we were ready to set off again, the snow on the road was up to our ankles. I called my father. In an incredibly soothing tone, he told me that sometimes the biggest act of courage is to know when to give up.

I owe him that lesson in caution, but also something more: its mathematical foundation.

Among his many obsessions, my father cares particularly about speed limits. Whenever a car would overtake us, zooming past on the motorway, he'd always say that the driver clearly didn't know that the violence of a crash doesn't increase proportionally to the speed of the car, but to its square value. I was a child, far away from the basic notions needed to make sense of that sentence. Years later, I looked back at it through physics: in the formula for kinetic energy (the energy of a body in motion), it's not velocity (v) that makes an appearance, but its square value:

$$E = \frac{1}{2}\ mv^2$$

Therefore, the crash was the energy (E), and my father was talking about the difference between a linear and a non-linear increase. He was preparing me for the fact that intuitive thought can sometimes be wrong. Going over the motorway speed limit wasn't more dangerous than I thought: it was *way* more dangerous.

# Hand, foot and mouth

In Milan, schools, universities, theatres, gyms are all closed. I keep receiving photos of the deserted city centre: it's 2 March but it's as quiet as *ferragosto*, the August bank holiday. Things are still relatively normal here in Rome, but it is a type of suspended, forced normality: the looming feeling that things are about to change is palpable everywhere.

The contagion has already compromised our bonds with the people around us and it has left loneliness in its wake: the loneliness of those in intensive care, communicating with other people through a pane of glass, but also a different kind of isolation, more diffused, more subtle, recognisable in the lips pressed behind surgical masks, in the suspicious looks, in having to be confined at home. In times of contagion we are all simultaneously free and under house arrest.

A week before I turned twelve, I contracted an illness called hand, foot and mouth disease. I was covered in itchy spots around my mouth and my extremities, as the name suggests. I had no fever, I didn't even feel ill other than for the itching, but I was extremely contagious and so I was placed in a form of quarantine. They gave me a pair of white cloth gloves to wear whenever I left my room, like the Invisible Man. Though it was nothing serious – just a silly, spotty disease – I remember I felt very alone, disheartened, and that I cried on my birthday.

No one likes being left out. And the knowledge that our separation from the world is only temporary doesn't do much to ease that suffering. We have a desperate need to be with others, among others, under three feet away from people who mean something to us. It feels as vital to us as breathing.

And so we feel the impulse to rebel: I won't let a virus stop my social interactions. Not for a month, not for a week, not even for a minute.

They're telling us we have to, but who is truly right here?

# The quarantine dilemma

Contagion, in its cold mathematical abstraction, is also a big game. A morbid game, but a game nonetheless, with its rules, its strategies, its goals (preserving who we are/not getting ill) and obviously its players – us. A game we could call 'the quarantine dilemma'.

Let's suppose that we're planning a friend's birthday party; it's happening tonight, though it's strange to have it on a Monday. The party is taking place in a really small club. However, the health minister – better, the World Health Organization – is telling people to avoid gatherings, to keep at a distance from coughing fits and sneezes. It's clear that there is no way to adhere to the safety guidelines on social distancing at a party. And even if we did follow the rules, can you imagine how sad it would be?

Each of us has two options: go to the party and hope for the best, or stay at home and sulk, thinking of everyone else having a blast. I know that all the guests are considering both options and a little mean streak in me is hoping that several will choose not to attend, so that I can still go and breathe a little more easily. But then I think about what would happen if everyone reached my same conclusion and showed up despite the warnings: what would happen if even just one of us happened to be infected and then . . . No, I don't want to go down that road.

Maths takes into account all the possible scenarios. With its predisposition to straight talking, it assigns numerical values to each guest's choice, arranges them tidily into a table and observes what happens moving from one cell to the next. Who loses, who wins. Then it comes back to us with a result that isn't as intuitive as we thought: the best choice isn't the one made just on the basis of my own personal gain. The best choice is the one that considers both my gain and – at the same time – that of all the people around me.

In brief: I'm sorry, but we'll have to reschedule.

# Against fatalism

And so the epidemic encourages us to think of ourselves as belonging to a collective. It pushes us to behave in a way that is unthinkable under normal circumstances, to recognise that we are inextricably connected to other people, to consider their exist-ence and wellbeing in our individual choices. In the contagion we rediscover ourselves as part of a single organism. In the contagion we become, again, a community.

Which brings us to an objection that was raised time and again in the early days of the epidemic: if the virus is as moderately lethal as reported – espe-cially for young, healthy individuals – why shouldn't I be free to take my life into my own hands and carry on with business as usual? Isn't a pinch of fatalism our inalienable right as free citizens?

The answer is no, we cannot put ourselves at risk. For at least two reasons.

The first is purely numerical. The percentage of necessary hospitalisations for Covid-19 is not negligible. Based on current estimates, which may very well change, around 10 per cent of the Infected end up in hospital. Too many contagions in a short time would mean 10 per cent of a much larger number, leading to too many patients and not enough beds or medical staff. So many, in fact, that the entire health system would break down.

The second reason is purely human. It pertains to that subsection of Susceptibles who are a little more susceptible than others: the elderly, people with pre-existing medical issues. Let's call them Ultrasusceptibles. If we, the young and healthy, expose ourselves to the virus more, we automatically bring it closer to them too. During an epidemic, Susceptibles must protect themselves in order to protect others. Susceptibles are also a buffer.

In times of contagion, therefore, what we do or don't do is no longer just about us. This is the one thing I wish for us never to forget, even after this is over.

So I look for a pithy formula, a memorable slogan, and I find it in an article published in *Science* in 1972: 'More Is Different'. When Philip Warren Anderson wrote it, he was referring to molecules, but he was also talking about us: the cumulative effect of our actions upon the collective is different

26

from the sum of the single effects. If there are many of us, each of our choices has global consequences, sometimes abstract and hard to understand: during a contagion, the lack of solidarity is first of all a lack of imagination.

# Against fatalism again

The community we should be taking care of isn't our neighbourhood or our city. It's not a region, or Italy, or even Europe. Community, in the contagion, is the entirety of the human species.

So, if we were patting ourselves on the back for our efforts to safeguard our national health service, we can stop that, right now. Here comes a more challenging thought: let's imagine what might – what *will* – happen if Covid-19 starts spreading with the same ease in parts of the African continent, where hospitals are in worse conditions than ours. Or where there is no health infrastructure at all.

In 2010 I visited a Doctors Without Borders mission in Kinshasa, in the Democratic Republic of the Congo. The mission was specifically working on HIV prevention and assistance to HIV-positive people, especially sex workers and their children. I still have a very vivid memory of the warehouse that served as a giant brothel, where families lived together, separated only by filthy rags, and mothers would sell themselves next to their children. I remember it well because it was the first time I was faced with such stark misery – beyond inhuman – and it shocked me to my core.

Now I try imagining the virus entering that warehouse, because we didn't do enough to contain it, because we really had to go to that birthday party. Who will have to face, then, the consequences of our privileged fatalism?

We are not all Susceptible in the same way, but Ultrasusceptibles are not only based on age and medical conditions: there are millions and millions of people who fall into the Ultrasusceptible category due to social and financial circumstances. Their fate, even if we see them as geographically distant, should feel very close to us.

# No man is an island

When I was in high school, people organised a lot of demonstrations against globalisation. I attended only one, and I was left a little disappointed. I couldn't understand what we were complaining about; everything felt too abstract, too generic. To be entirely honest, I quite liked the idea of globalisation: to me it meant a lot of good music and fun adventures abroad.

To this day, the term 'globalisation' leaves me a little confused. It's still a many-faced creature with no clearly defined nature, but now I can at least see its outline, mapped out by its collateral effects. Such as a pandemic, for example. Or this new form of shared responsibility no one can ignore any longer.

No one. If interactions between human beings were drawn in pen, the world would be a giant scribble. Even the most rigorous of hermits, in 2020, has their minimum quota of connections. We live in an extremely connected graph, to use mathematical terms. The virus runs along the pen lines and reaches everywhere.

In times of contagion, the much-abused John Donne reflection, 'No man is an island', takes on a new, darker meaning.

# Flying

We are not marbles. We are human beings, filled with dreams and neuroses. We are, most of all, filled with commitments. We travel more frequently and further afield than any past generation, and we exchange words and goods with such a large number of people that it would leave our ancestors baffled.

If we're harbouring a nasty cold, the virus moves with us, inside us, and is scattered a little here, a little there, in Milan, in London, at the supermarket where we shop every few days, at our parents' place where we had lunch last Sunday. The contagion is impartial, especially if it happens through sneezes, and is even more effective if the majority of the Infected remain asymptomatic. Just like bees carry pollen around, so we carry our anxieties and our pathogens.

In 2002, Sars-Cov made its debut in a market in Guangdong, a province in the south of China. A doctor was infected while in the hospital and carried the virus to a Hong Kong hotel. Two women contracted the virus in the hotel and travelled to Toronto and Singapore, where other clusters appeared. Following other routes, the contagion touched Europe too, with limited consequences at the time.

Flight traffic has changed the fate of viruses, allowing them to colonise even more land even further away. But it's not only thanks to flights. There are trains and buses, cars and these new electric scooters. The simultaneous movement of 7.5 billion people: that is the coronavirus transport system. Fast, comfortable, efficient – just the way we like it.

In times of contagion, our efficiency is also our downfall.

# Chaos

All of those movements, taken together, signify chaos on a gargantuan scale. The word *chaos* suggests the idea of something that goes beyond maths' remit, beyond rationality itself. That is not the case. There are very refined and efficient techniques to control it; there are equations, or rather clusters of equations, linked to each other in order to observe how a chaotic system will evolve in the future.

Weather forecasting follows these techniques. Meteorologists gather measurements from an enormous number of thermometers and barometers across the globe, along with satellite images, wind speeds and precipitations, and use this vast amount of data to feed the equations of atmospheric models. They load their simulations onto computers and receive tomorrow's weather, with its probability attached.

Today, however, is 3 March 2020 and we're dealing with a different kind of forecast. We need a lot of data, more than we can express. We need to know how many people live on each tiny patch of land and all their comings and goings. We need everyone's movements, and that's still not enough. We know that the epidemic changes if we change – if we stop going to work, if we keep our distance, if we are afraid, or very afraid – and our simulations need to consider all that too.

Mathematicians are working on this – and with them physicists, doctors, epidemiologists, sociologists, psychologists, anthropologists, urbanologists, climatologists. Scientists have never slept so little. Everyone is trying to fill in the SIR model to see where Cov-2 might head tomorrow. If we are successful in our simulation, we will have gained a few days' lead.

# At the market

We know more about the future of Cov-2 than about its past. The circumstances surrounding its birth aren't clear and it might still be some time before we discover them. The general mechanics, however, are familiar: Cov-2 – just like Sars and AIDS – has infected humans through another animal species.

Everyone is pointing a finger at bats, who also carried Sars. But Cov-2 wasn't transmitted directly from bats to humans; it stopped somewhere else, in another species, perhaps a snake. Inside this new host, its RNA mutated to the point where it became dangerous to us. At that point, it made its second leap, and infected one or more humans, the patients zero of this global event.

We think that all this happened in China, in a market in Wuhan, where different species of wild animals are kept within close proximity, an extremely favourable environment for the spread of pathogens. Tracing exactly how, when and where the first leap happened isn't just curiosity for its own sake, but a mission for epidemiology, as important as containing the virus. It will, though, be much slower, and even more difficult.

And, yet, too many people have already summarised Cov-2's history in a few, harsh words: 'People in China eat disgusting animals. And they eat them alive.'

# At the supermarket

I have a friend who married a Japanese woman. They live close to Milan and have a five-year-old daughter. Yesterday, mother and daughter were in the supermarket and a couple of men started yelling at them, saying it was all their fault, they should go home, back to China.

Fear makes us do strange things. In 1982, the year I was born, Italy reported its first case of AIDS. My father was a 34-year-old surgeon at the time. He tells me that, in that initial phase, neither he nor his colleagues knew what to do; no one had any idea what the virus might be. When they needed to operate on an infected patient, they wore two pairs of gloves as a precaution. One day during surgery, a drop of blood from an HIV-positive woman's arm fell onto the floor of the operating room and the anaesthetist jumped back, screaming.

They were all doctors, and they were all terrified. No one is ever ready for a task that is completely new. In circumstances like those we're currently living through, every form of human reaction makes its appearance: anger, panic, indifference, cynicism, disbelief, resignation. If only we would remember this and then try to be a little more careful than usual, a little kinder. Then perhaps we wouldn't start hurling insults in supermarket aisles.

And even without considering people's inability to differentiate between distinct Asian traits, we should really understand that 'they' are not to blame for the contagion.

We are.

# Moving

The world is still a wonderfully wild place. We think we have explored all of it, but there are still microbial universes about which we know nothing, interactions between species that we haven't even begun to imagine.

Our aggressive behaviour towards the environment increases the likelihood of coming into contact with these new pathogens, which until now were happily confined to their natural niches.

Deforestation brings us closer to habitats that never considered our presence; our unstoppable urbanisation does the same.

The accelerated extinction of various animal species is forcing several bacteria that lived inside their guts to move elsewhere.

Intensive farming creates involuntary cultures where literally anything proliferates.

Who among us can truly know what last summer's Amazon forest fires have set free? Who is able to predict what will be the aftermath of the almost mass extinction that recently took place in Australia? Micro-organisms that science has never even named might soon be in need of a new home. And we are the perfect breeding ground: there are so many of us – and there will be so many more – we are so Susceptible, have so many connections and we move around so much.

# Too easy a prophecy

Viruses are among the many refugees of environmental destruction. Right next to bacteria, fungi, protozoa. If we were able to set aside some of our egocentrism, we'd realise that it's not these new microbes seeking us out, but rather our actions that are unearthing them.

The growing need for food is forcing millions of people to resort to eating animals that should be left alone. In West African countries, for example, there has been an increase in the consumption of bushmeat, including bats, who are among the unfortunate carriers of Ebola in the region.

Contact between bats and gorillas – from whom Ebola can then easily be contracted by humans – is made more likely due to the overabundance of mature fruits on trees, which is in turn due to the increasingly drastic succession of freak downpours and dry seasons, in turn brought about by climate change . . .

It's almost too much to bear. A catastrophic sequence of causes and effects. But these chain reactions – of which there are many – need to be urgently considered and understood by all of us. Because at their end there might be another pandemic, even worse than the one we are experiencing today; and because their origin, as remote as it may seem, resides always and inevitably with us and our destructive behaviours.

I allowed myself some emphasis, at the start, when I said that what is happening has already happened before and will happen again. It wasn't a prophecy. It wasn't even guesswork. In fact, I can easily add now that what is happening with Covid-19 will keep happening more often.

The contagion is just a symptom. The infection is in our ecosystem.

# Hairspray

In the 1980s, big hair was all the rage. Hundreds of tonnes of hairspray were released into the atmosphere every day. Then it turned out that CFCs were opening a hole in the ozone layer, and that if we didn't do something about it, the sun would roast us all alive. Everyone changed their hairstyle and humanity was saved.

Back then, we acted with cooperation and efficiency. But the hole in the ozone was easy to imagine; it was a hole, after all, and all of us are able to imagine a hole. What we are being asked to visualise and understand right now is a lot more abstract.

This is the paradox of our times: as reality becomes more complex, we become more resistant to complexity itself.

Take, for example, climate change. The increase in global temperatures is tied to geopolitical scenarios linked to petrol prices *and* to our own summer holiday plans, to switching off lights in the corridor *and* to economic competition between China and the United States; to the meat we buy at our local market *and* reckless deforestation. Personal and global are intertwined in such enigmatic ways that they leave us exhausted even before we can try following a single thread.

It's even worse when we move from the causes to the consequences of global warming: on one side the Amazon fires, on the other torrential rain in Indonesia; the hottest summer of the century but also the coldest winter. Scientists keep warning us that we might not survive, but they also tell us that our own perception of a particularly muggy summer's day is irrelevant, that one day is not statistically significant and one person complaining about the hot weather is even less so.

The only certainty, it seems, is that our brain is not equipped to deal with this. But we'd do well to start prepping for change: among the numerous illnesses that could benefit from climate change are, along with Ebola, malaria, dengue fever, cholera, Lyme disease, the West Nile virus and even diarrhoea, which may be little more than an inconvenience in wealthier societies, but poses a serious threat in some countries.

The world is about to shit itself.

The contagion, then, is an invitation to think. And quarantine is the opportunity to do so, a chance to find the time to understand that we're not only part of the human species: we are members of the most invasive species of a fragile, magnificent ecosystem.

# Parasites

I spend my summers in Puglia, in the south of Italy. Whenever I think about that place, from afar, the first thing that comes to mind is the olive trees. On the road that leads from Ostuni to the sea, there are specimens so ancient and majestic that you can't help but think of them as more than just plants. Their bark is so expressive they appear to be sentient. Sometimes, I'll admit it, I have given in to the magical impulse to hug one, to try to steal some of its power.

The olive tree parasite *Xylella fastidiosa* was first detected in the fields near the town of Gallipoli in 2010. From there, it started its slow, patient march northwards, infecting olive groves across the region, mile by mile. Initially it looked like nothing more than a few wisps of sunburnt leaves, but, with time, whole trees turned into skeletons. Last summer, driving on the main road from Brindisi to Lecce, I was surrounded by cemeteries of grey wood.

Nevertheless, ten years have not been enough to bring everyone to agree.

Xylella exists.

No, Xylella doesn't exist.

Xylella will infect all of the olive trees.

Xylella only infects trees that are not taken care of.

Xylella is caused by weedkillers.

Xylella comes from China (it's their fault).

We need to uproot every tree in a 300ft radius from an infected one.

All you need is the true and tested lime on the bark. Hands off the olive trees!

The epidemic is a regional issue.

The epidemic is a national issue.

The epidemic is a European issue.

Meanwhile, the parasite continued its march, and reproduced in peace. It showed up in Antibes, in Corsica, in Majorca. Xylella apparently loves holiday spots.

# Experts

March 4th. The Italian government has announced the closure of all schools across the country and I've already argued with a number of people. In these times of contagion, the main argument concerns the difference between Covid-19 and seasonal flu. But there are also heated debates about containment measures, which are either too lax or too strict.

It has always been like this, since it started: on one side, there are those who point out the fact that infected people end up in hospital; on the other, those who talk about it as if it's nothing more than an overhyped cold. Those who say that you just need to wash your hands a little longer, a little more carefully, and those who ask for the entire country to be put under quarantine.

'Experts say . . .'

'Let's hear some experts . . .'

'But experts are saying that . . .'

'What is sacred in science is truth,' wrote Simone Weil. But what is truth, when we're testing the same data, sharing the same models and then reaching different conclusions?

In the contagion, science has disappointed us. We wanted certainties and we were given opinions. We forgot that this is how it always works, that it can only work like this, that for science doubt is even more sacred than truth. Right now, we don't care. We're looking at experts bickering among themselves like toddlers watching their parents fight.

And then we start bickering too.

# Foreign multinational corporations

Where there is no agreement in science, suppositions, half-truths or full-blown lies arise.

Xylella is lab-created, developed by foreign multinational corporations to undermine our national olive oil production industry.

Actually, it was made to build golf courses across Puglia.

Climate change is part of a natural cycle.

Greta Thunberg is paid by foreign multinationals and keeps wasting plastic wherever she goes.

Coronavirus is also a lab creation, developed by foreign multinationals to profit from the sale of its vaccine.

Yet another vaccine that will cause autism in children.

Seasonal flu causes more deaths than Covid-19.

The Chinese knew about all this, anyway.

The Americans knew about all this, anyway.

Bill Gates knew about all this, anyway.

People are shooting each other in the streets of Wuhan, right now.

We are free to believe that Cov-2 has spread among the Chinese from a vial smuggled out of a lab where secret military experiments were taking place. It's probably more fascinating than bat transmission. It's a theory, however, that demands many more arbitrary assumptions – the existence of the lab, of the military project, of the vial and of a plot to steal it – compared to a documented phenomenon that has been proven to have happened multiple times in the past.

In cases like this, science refers to Occam's razor – that is, always take the shortcut. The simplest solution, the one that requires fewer leaps of logic, is most likely to be the correct one.

I know the theory of the secret lab is more fun but let's leave it where it belongs: as a subplot in the next Bond movie.

# The Great Wall

For twenty years, I believed that the Great Wall was the only human construction visible from space. I believed it because that is what people said and we can often believe in things if we don't think about them too much. When I finally stepped onto the wall, I realised that it made no sense: it's an impressive construction, but it's also quite narrow. There is no way it could be seen from up there.

False information spreads just like epidemics: the model used to study its diffusion is the same. When faced with a piece of false information, we are either Susceptible, Infected or Recovered. And the scarier the information – the more provocative, the more outrageous – the more vulnerable we are to the contagion.

Yesterday, the big news making the rounds on social media was that the epidemic was slowing down in Italy. Today, experts have been working around the clock to prove that it's not true: there is no proof, not yet. The news, however, was endemic. It was on Facebook, on Twitter, in all of our multiple WhatsApp groups. Just as Covid-19 moves around on planes, lies spread incredibly quickly through smartphones.

Inevitably, people will be disappointed that there is no slowing down. Eventually, their disappointment will generate more suppositions on the reason why the contagion is not coming to a halt. Those suppositions will be added to the previous assumptions. And so on.

Even half-formed suppositions are part of an ecosystem, a boundless one in which anything can happen.

# The god Pan

When the Italian newspapers made the decision to stop putting on their front pages the numbers of the contagion, I felt annoyed and betrayed. I started looking for the information in other sources. In times of contagion, transparent information isn't a right: it's a crucial preventative measure.

The more a Susceptible individual is informed – about numbers, places, patient concentrations in hospitals – the more their behaviour will adjust to the context. Of course, this is not always the case – there will always be someone who reacts unpredictably – but most of us are equipped with reasoning skills. Scientific simulations take into account our awareness as a cushioning factor for the epidemic.

And yet, since the very first day, numbers have been blamed for causing panic. So it's better to hide them, or to find a different way of counting, a way that would make the numbers seem smaller. Except we then realised almost immediately that the direct consequence of this new strategy was, in fact, *actual* panic: if the truth is being hidden from us, then everything must be way more serious than what they want us to believe. After a couple of days, the numbers were back on all the homepages.

This confusion is the sign of an unresolved relationship, a love triangle that seems to have come undone in our modern times: a relationship drama where citizens, institutions and experts seem unable to communicate.

While institutions might trust experts, they seem wary of us, of our emotional stability. Not even experts, truth be told, have much faith in the people: they talk to us in too simple a manner, causing suspicion. We, the people, have always been sceptical of institutions – we were so before and we always will be – which then makes us turn to the experts, only to see them hesitate. In the end, the uncertainty is what makes us behave in even worse a manner than we ordinarily would, causing the cycle of suspicion to start all over again.

The virus has focused a spotlight on this vicious circle, a loop of distrust that arises almost every time that science brushes against our daily lives. It's this loop, and not the numbers, that generates panic.

After all, the ancient Greeks believed panic to be a circular invention of the god Pan: sometimes he would emit cries so loud and terrifying that he would scare himself with his own voice and try running away from himself in fear.

# Numbering our days

I just received an email. I should have been attending a conference in Zagreb. The idea was to bring together representatives of various disciplines and countries to try to work out a new definition of what it means to be European. The organisers are now asking me to 'reconsider my participation'. Authorities are recommending avoiding the presence of guests from high-risk areas, and Italy is among them, along with China, Singapore, Japan, Hong Kong, South Korea and Iran. A curious gang. The G7 of contagion.

As the epidemic continues, nearing 100,000 cases, I bear witness to the crumbling of my calendar. March will be very different from expected. April, who knows. It's a strange feeling, this loss of control; I'm not used to it, but I'm not resisting it either. None of these missed opportunities can't be rescheduled, or lost and made peace with, with no regrets. We are facing something much bigger that deserves our focus and respect.

A lot of this crisis has to do with time. With our way of organising, squeezing time. But now that we are at the mercy of this new microscopic force that has the audacity to decide for us, we find ourselves under pressure, angry, as if stuck in a traffic jam but with no one else around us. In the invisible grip of the virus, we yearn for a return to normal life; we feel like we have *the right* to go back to normal. All of a sudden, normality is the most sacred thing we have – even though we had never given it so much importance before, even though we don't actually know what it is. We just know we want it back.

Normality, however, has been suspended, and no one can foresee for how long. This is the time of anomaly; we need to learn to live with it, in it. We need to find reasons to welcome it, other than just our fear of death. It may be true that viruses have no intelligence, but they are better than us at this: they change, they adapt, and they do so quickly. We should learn from them.

Our current stalemate will have immeasurable consequences – jobs will be lost, shutters closed; every sector will be impacted; everyone is already dealing with their own losses. Our civilisation can afford anything except slowing down.

But what will happen *after* is a thought too complex for me; I can't grasp it, I give up every time I try. I will take on the new realities once they arrive, one at a time.

In Psalm 90 there is a line that I often think about these days:

> *Teach us to number our days,*
> *that we may gain a heart of wisdom.*

Maybe it comes to mind because all we can focus on during this epidemic is numbers. The number of infections and recoveries; the number of deceased; the number of hospitalisations and missed school days; the numbers of the billions lost on the stock markets, of masks sold and of hours left until the results of the test; the number of miles from the nearest cluster and of cancelled hotel-room bookings; the number of our ties, our contacts, our sacrifices. And we're counting and counting again the number of days, especially the days that are left until the emergency is over.

I have the feeling, however, that the Psalm is suggesting a different kind of numbering: teach us to number our days so we may give value to our days. *All* of our days, even those belonging to this painful interval.

We can tell each other that Covid-19 is an isolated incident, a calamity or a scourge, cry that it's all 'their' fault. We're free to do so. Or we could try making sense of the contagion.

Make better use of this time, use it to think about what our busy normality prevents us from considering: how did we get here, how do we want to start again?

Number the days.

Gain a heart of wisdom.

Don't allow all of this suffering to be in vain.